Black Holes

Lily Erlic

LET'S READ

AV² BY WEIGL™

ADDED VALUE • AUDIO VISUAL

Go to **www.av2books.com**, and enter this book's unique code.

BOOK CODE

AVJ29757

AV² by Weigl brings you media enhanced books that support active learning.

AV² provides enriched content that supplements and complements this book. Weigl's AV² books strive to create inspired learning and engage young minds in a total learning experience.

Your AV² Media Enhanced books come alive with...

Audio
Listen to sections of the book read aloud.

Video
Watch informative video clips.

Embedded Weblinks
Gain additional information for research.

Try This!
Complete activities and hands-on experiments.

Key Words
Study vocabulary, and complete a matching word activity.

Quizzes
Test your knowledge.

Slideshow
View images and captions, and prepare a presentation.

... and much, much more!

Published by AV² by Weigl
350 5th Avenue, 59th Floor
New York, NY 10118
Website: www.av2books.com

Library of Congress Control Number: 2019941849

ISBN 978-1-7911-0966-0 (hardcover)
ISBN 978-1-7911-0967-7 (softcover)
ISBN 978-1-7911-0968-4 (multi-user eBook)

Printed in Guangzhou, China
1 2 3 4 5 6 7 8 9 0 23 22 21 20 19

052019
102918

Project Coordinator: John Willis
Designer: Terry Paulhus

Black Holes

CONTENTS

4

Black holes are places in space. They work like vacuums. Black holes pull things inside themselves.

Things pulled into black holes stretch out like spaghetti. Scientists call this the "noodle effect."

8

People cannot see black holes. However, they can see things being pulled into them. That is how scientists find black holes in space.

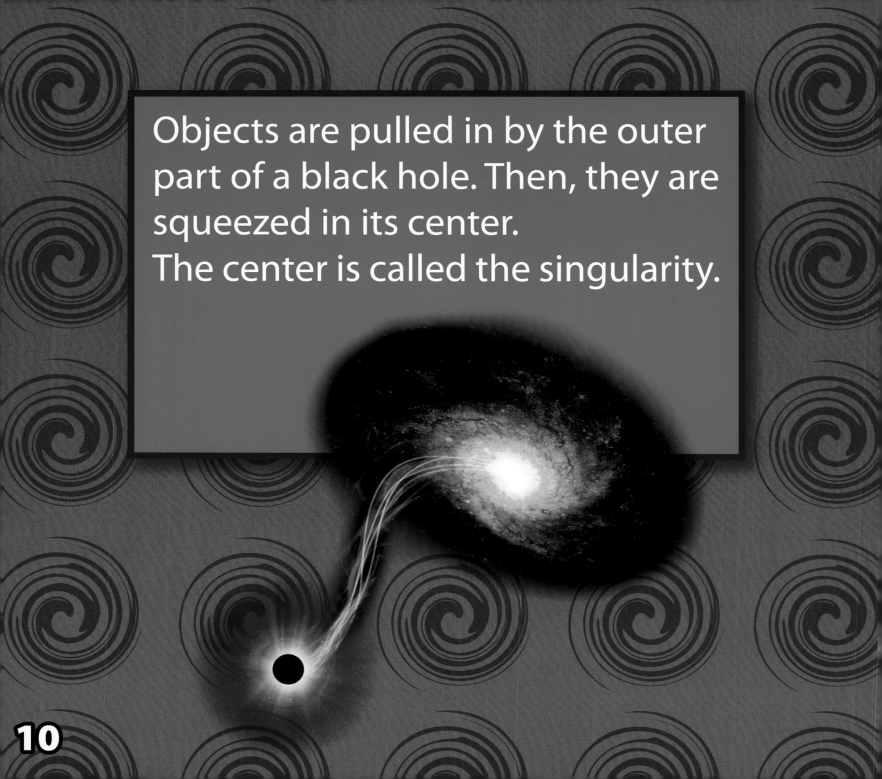

Objects are pulled in by the outer part of a black hole. Then, they are squeezed in its center.
The center is called the singularity.

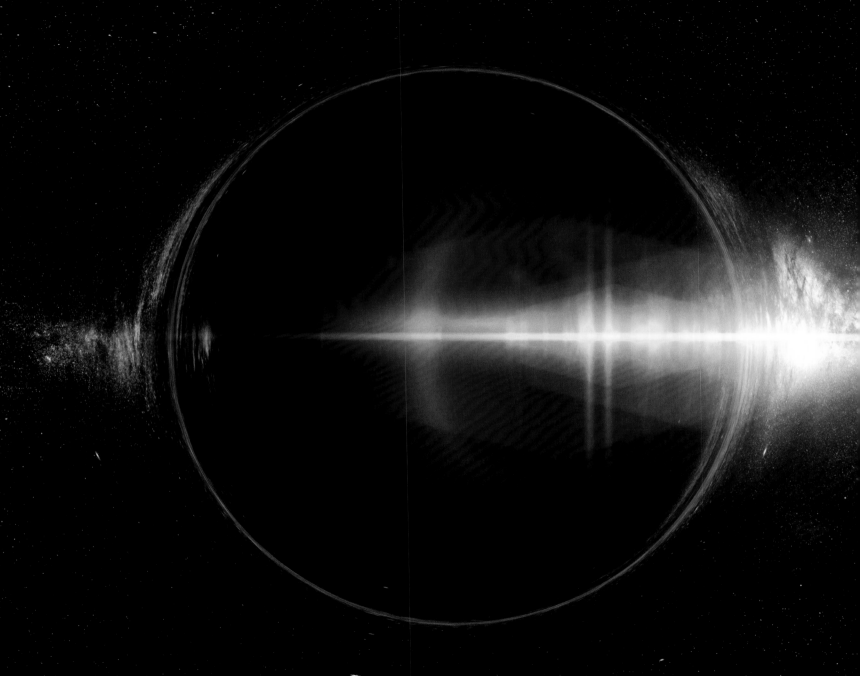

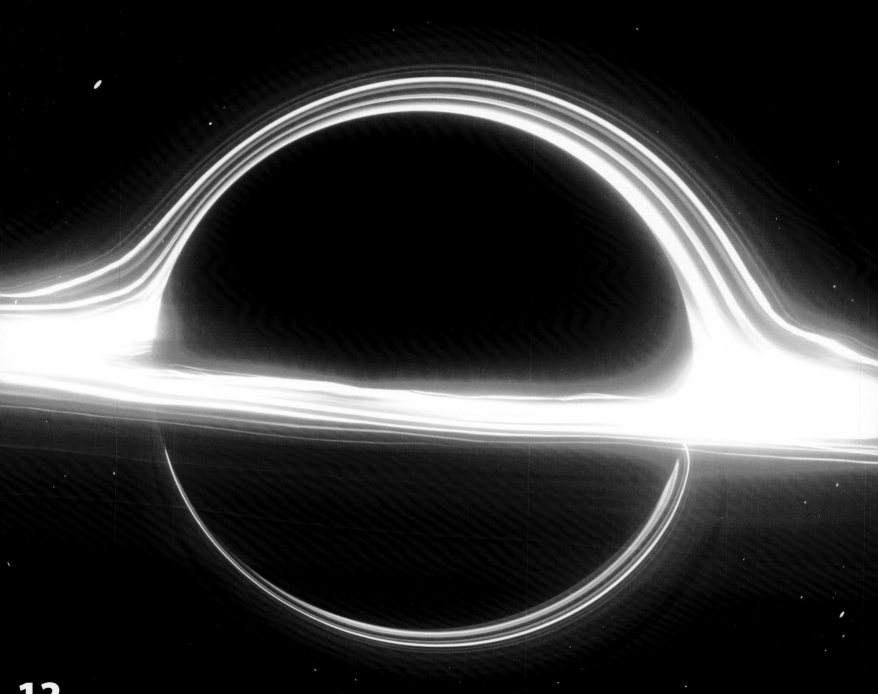

There are different kinds of black holes.
Miniature black holes are tiny.

Stellar black holes come from exploding stars. They are bigger than the Sun.

15

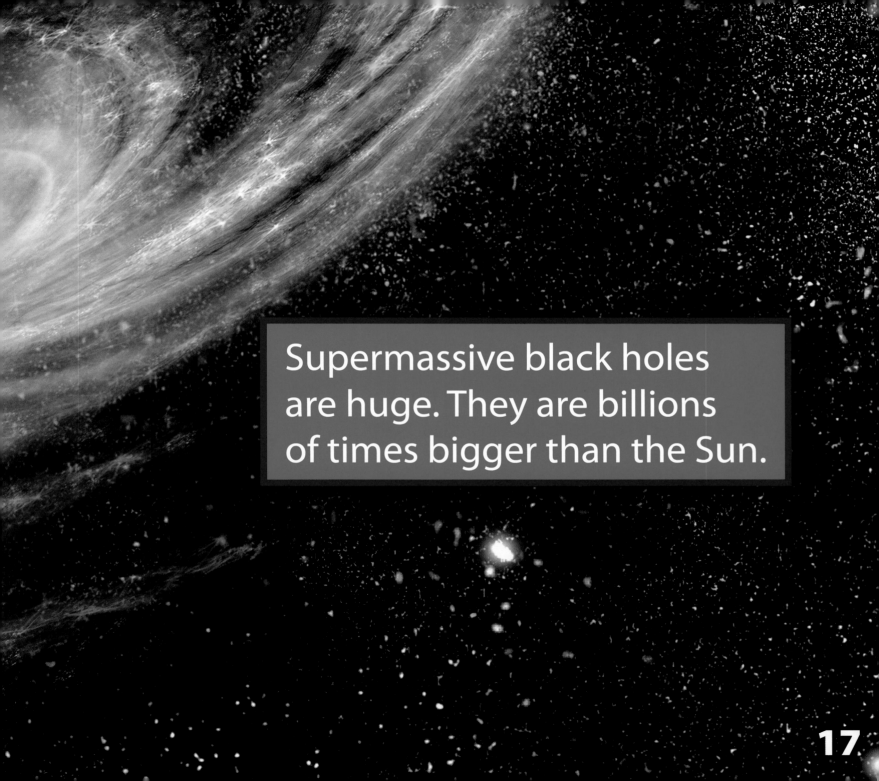

Supermassive black holes are huge. They are billions of times bigger than the Sun.

A galaxy is a group of stars and planets. Scientists think there is a supermassive black hole in the center of almost every galaxy.

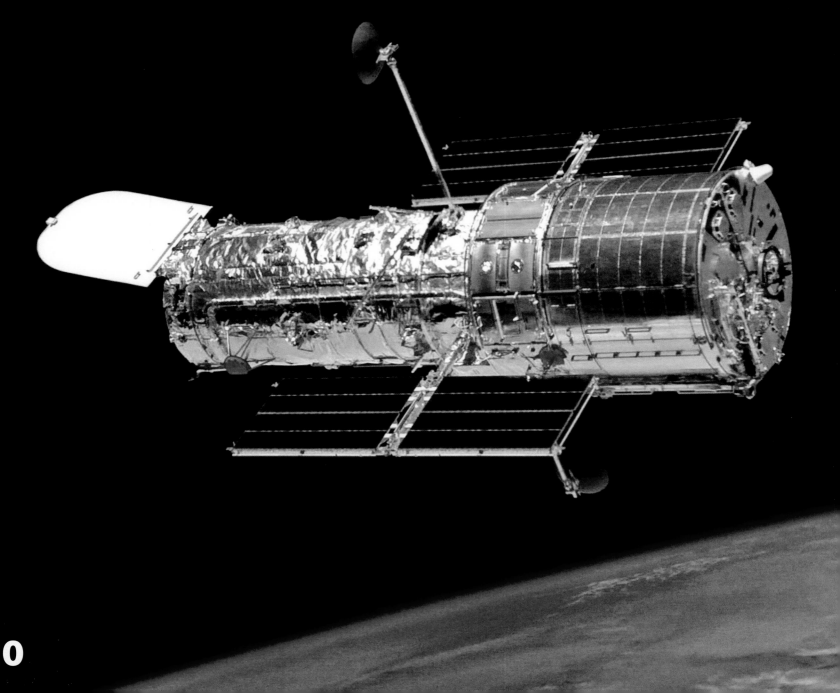

Scientists take pictures of objects around black holes using space telescopes. These pictures help them learn more about black holes.

BLACK HOLE FACTS

These pages provide detailed information that expands on the interesting facts found in the book. They are intended to be used by adults as a learning support to help young readers round out their knowledge of each object or event featured in the *Deep in Space* series.

Pages 4–5

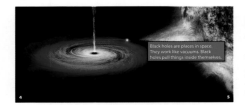

Black holes are places in space. They pull objects, such as stars, inside themselves. If stars get too close to black holes, they are pulled in by gravity. Gravity is a force that pushes or pulls an object. It helps people stay on the ground. Just like a vacuum pulling in crumbs, black holes pull in stars and other space objects.

Pages 6–7

Things pulled into black holes stretch out like spaghetti. If a star or planet travels near a black hole, it gets squeezed into a long, thin shape. Scientists can see stars stretch out like noodles in the black hole of the Milky Way Galaxy. This is called spaghettification, or the noodle effect. Once the space object begins to stretch into the black hole, it cannot escape.

Pages 8–9

People cannot see black holes. Black holes are invisible. Scientists know black holes are present by observing what happens outside of them. If stars are spinning around a dark spot in space, it is probably a black hole. Scientists also observe matter being pulled into black holes. They use special x-ray telescopes to study them.

Pages 10–11

Things get pulled in by the outer part of a black hole. The event horizon surrounds the black hole. There is gravity in the event horizon. Any space objects that come near the event horizon can be pulled into the black hole. The singularity is the center of a black hole where space objects have been squeezed together. Gravity is the strongest in the singularity.

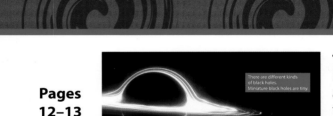
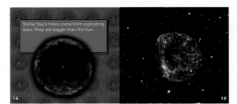
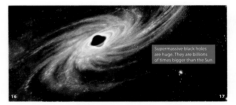

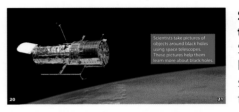

KEY WORDS

Research has shown that as much as 65 percent of all written material published in English is made up of 300 words. These 300 words cannot be taught using pictures or learned by sounding them out. They must be recognized by sight. This book contains 46 common sight words to help young readers improve their reading fluency and comprehension. This book also teaches young readers several important content words, such as proper nouns. These words are paired with pictures to aid in learning and improve understanding.

Page	Sight Words First Appearance
5	are, in, like, places, they, things, work
6	call, into, out, this
9	being, can, find, how, is, its, people, see, that, them
10	a, by, of, part, then
13	different, kinds, there
14	come, from, than
17	times
18	and, almost, every, group, think
21	about, around, help, learn, more, pictures, take, these

Page	Content Words First Appearance
5	black holes, space, vacuums
6	noodle effect, scientists, spaghetti
10	center, singularity
13	miniature black holes
14	stars, stellar black holes, Sun
17	supermassive black holes
18	galaxy, planets
21	space telescopes